D1324032

ANATOMY
Coloring Book for Students & Even Adults

The Anatomy Coloring book and Physiology Workbook with Magnificent Learning Structure

Enabling the students and individuals of different kinds to take the most out of their interest for anatomy, anatomy and physiology coloring book comes with thoroughly amazing structure. Netters anatomy coloring book is one of the most unique books of its own kind which covers colored human anatomy.

Features :

Anatomy and physiology coloring workbook is a tremendous solution for such ones who want to enhance their understanding about human anatomy and that's too in a colorful fashion.

Human anatomy coloring books features multiple illustrations, views and dissection layers which provide the users with varied opportunities to learn and enjoy flawlessly.

Anatomy coloring book for kids comes with such gradually advancing challenges and tasks so that the users could boost up their level of knowledge and education in a smooth way.

Anatomy coloring book has numerous hints and tracking techniques for coloring in such a way that it gets you along quite sensationally so that the users can have the best out of its abilities.

Patrick N. Peerson
Funny Learn Play

Table of Contents

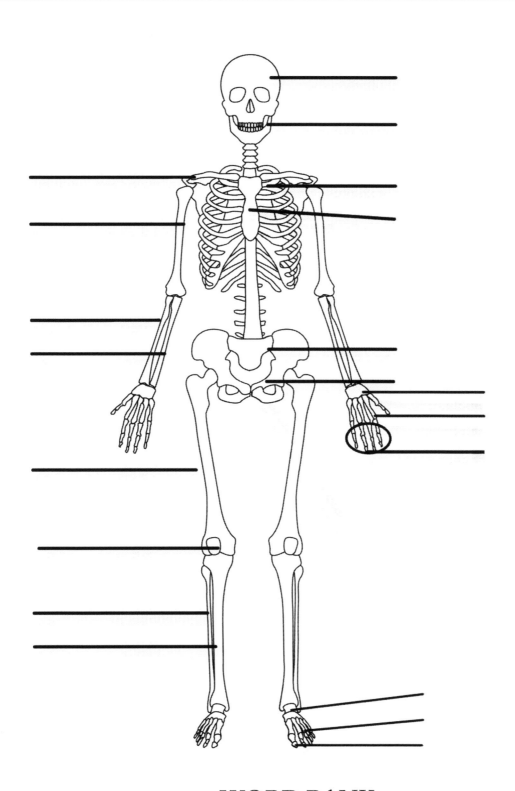

WORD BANK

Clavicle	Tarsals	Sacrum	Cranium	Femur
Humerus	Metatarsals	Carpals	Mandible	Patella
Radius	Phalanges	Metacarpals	Sternum	Tibia
Ulna	Pubis	Phalanges	Ribs	Fibula

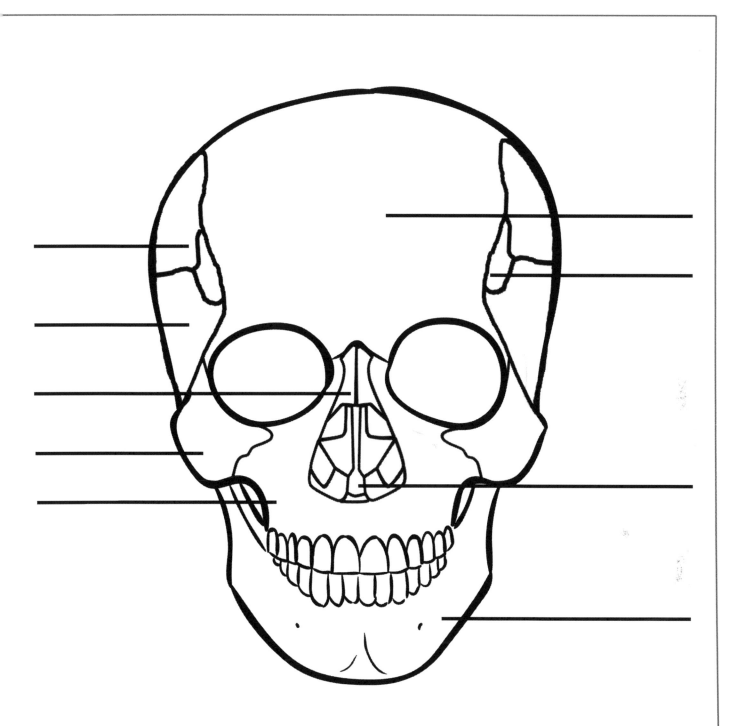

WORD BANK

Frontal bone
Parietal bone
Sphenoid bone

Temporal bone
Nasal bone
Zygomatic bone

Vomer
Maxilla
Mandible

5

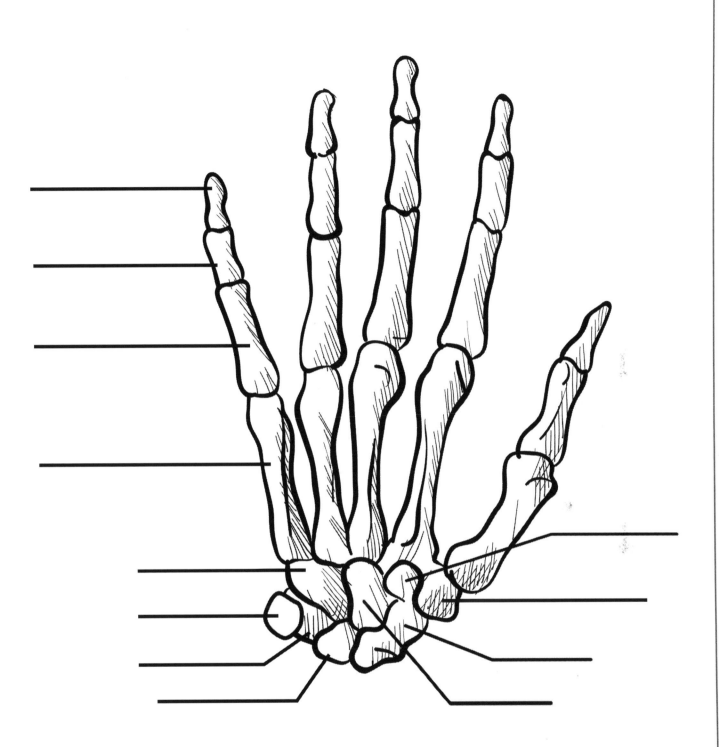

WORD BANK

Distal Phalanx

Middle Phalanx

Trapezium

Scaphoid

Triquetral

Lunate

Trapezoid

Proximal Phalanx

Metacarpal

Hamate

Pisiform

Capitate

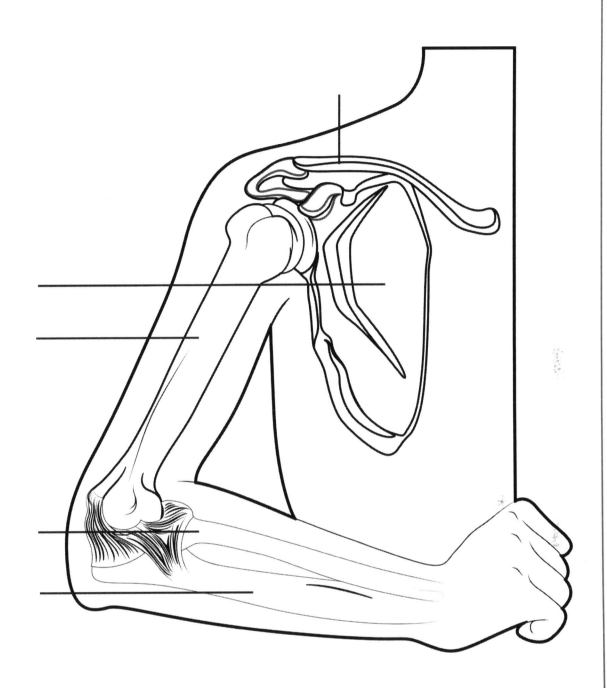

WORD BANK

Clavicle   Humerus   Radius
Scapula   Ulna

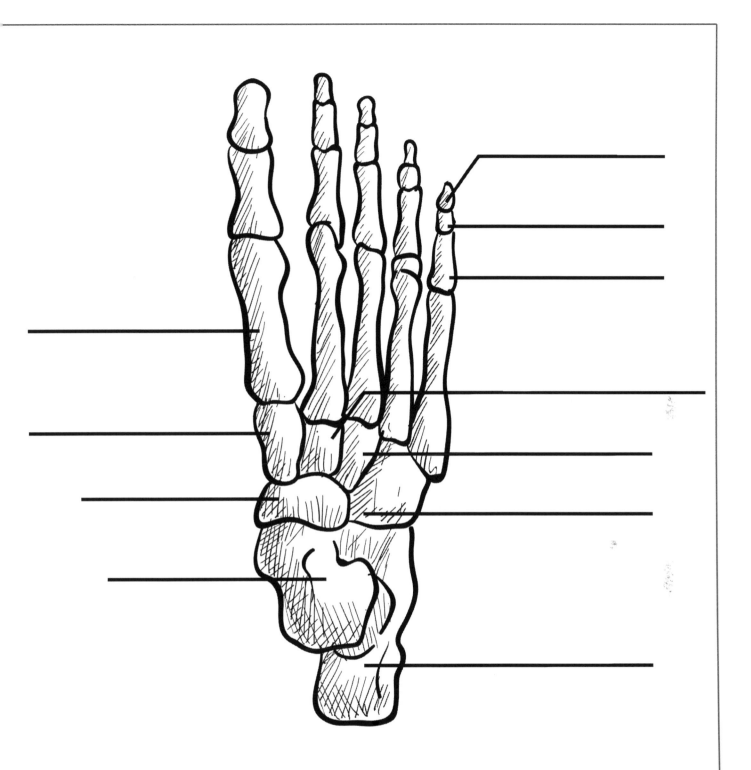

WORD BANK

Distal Phalanx   Medial cuneiform   Calcaneus
Middle Phalanx   Intermediate cuneiform   Cuboid
Proximal Phalanx   Lateral cuneiform   Navicular
Metatarsal   Talus

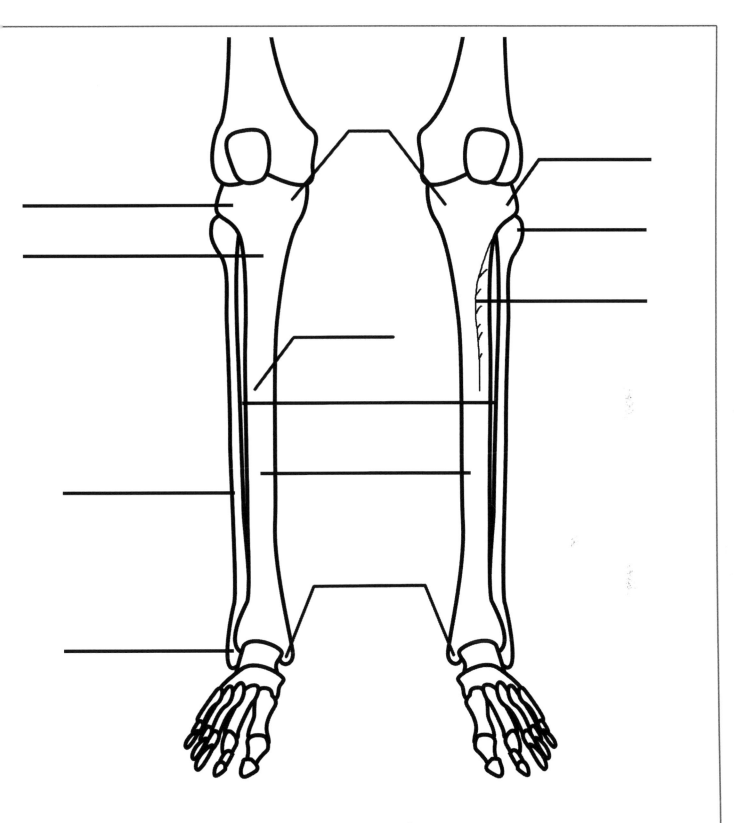

WORD BANK

Lateral condyle
Medial condyle
Anterior border
Interosseous membrane

Fibula
Lateral malleolus
Tibia
Soleal Line

Articular surface
Head of Fibula
Tibial tuberosity
Lateral condyle

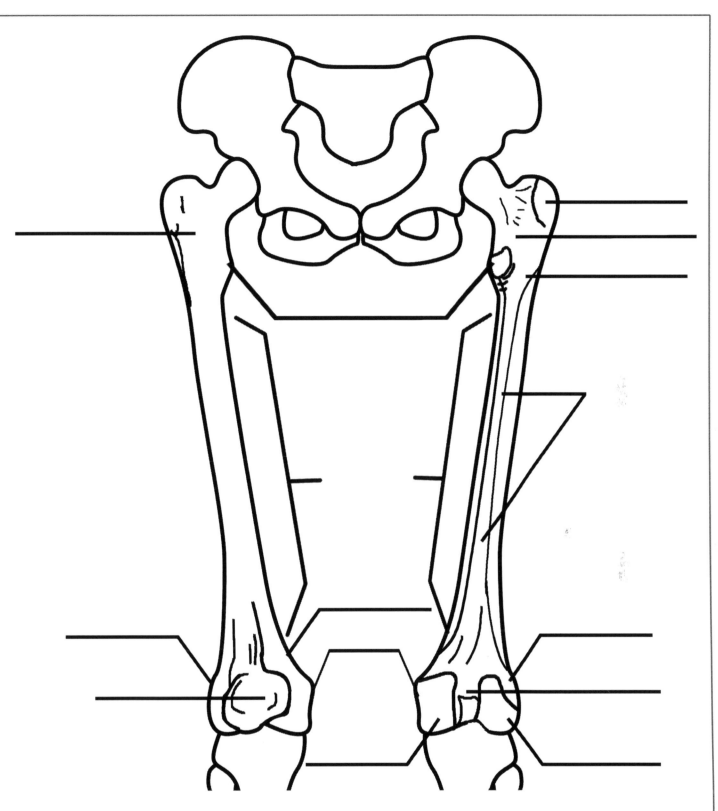

WORD BANK

Lateral epicondyle	Lateral condyle	Greater trochanter
Patella	Intercondylar fossa	Lesser trochanter
Medial epicondyle	Lateral epicondyle	Intertrochanteric crest
Adductor tubercle	Linea aspera	Intertrochanteric line
Medial condyle	Gluteal tuberosity	Body (shaft)

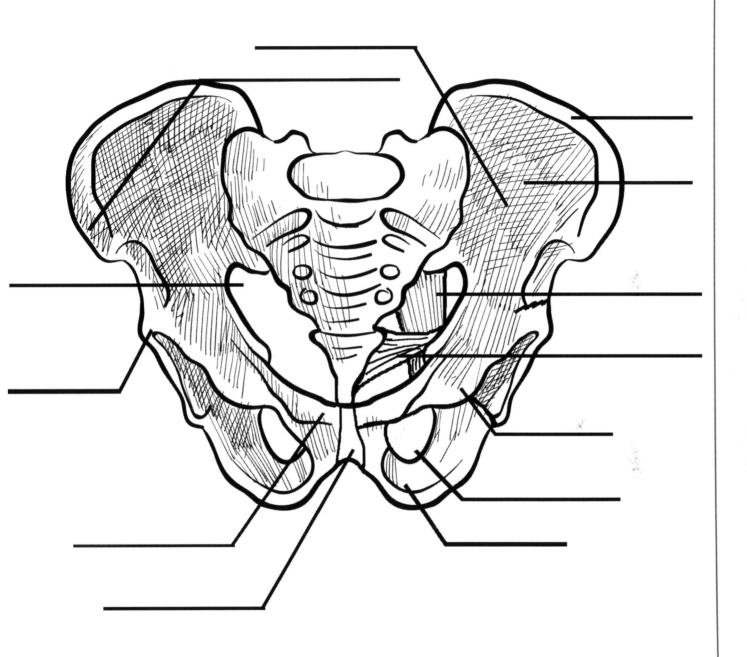

17

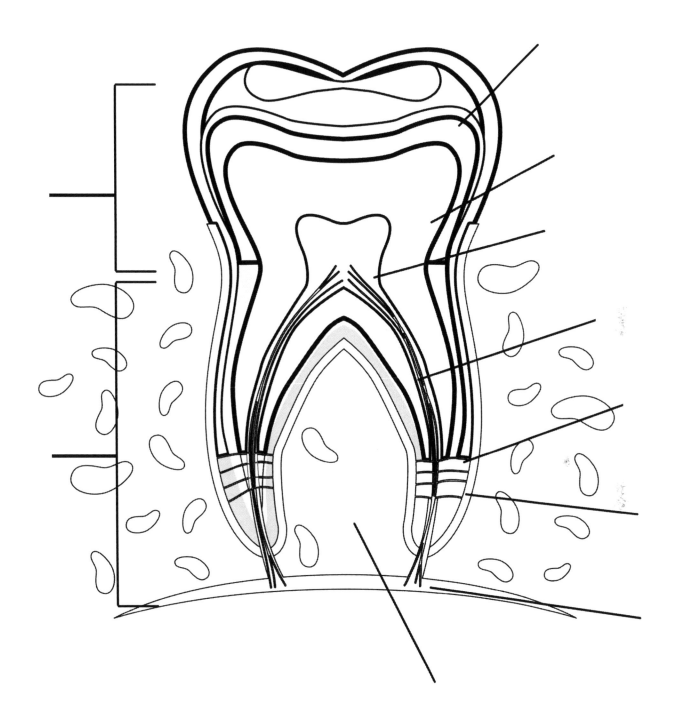

WORD BANK

Bone	Dentin	Root
Accesory canal	Enamel	Peridontal ligament
Root-End opening	Pulp chamber	Crown
Root canal containing pulp tissue		

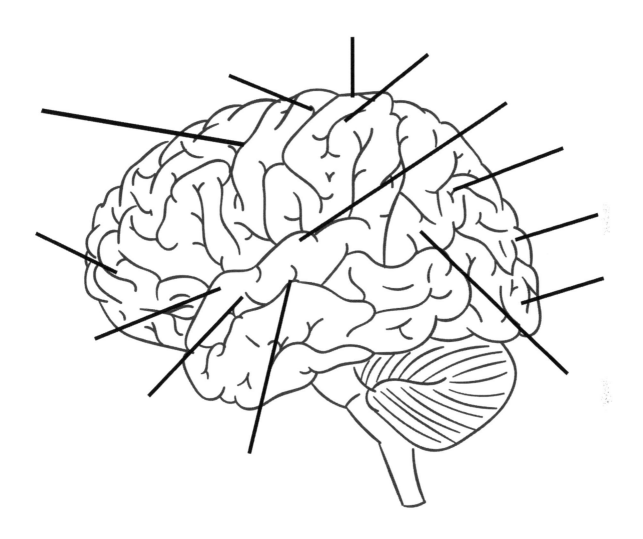

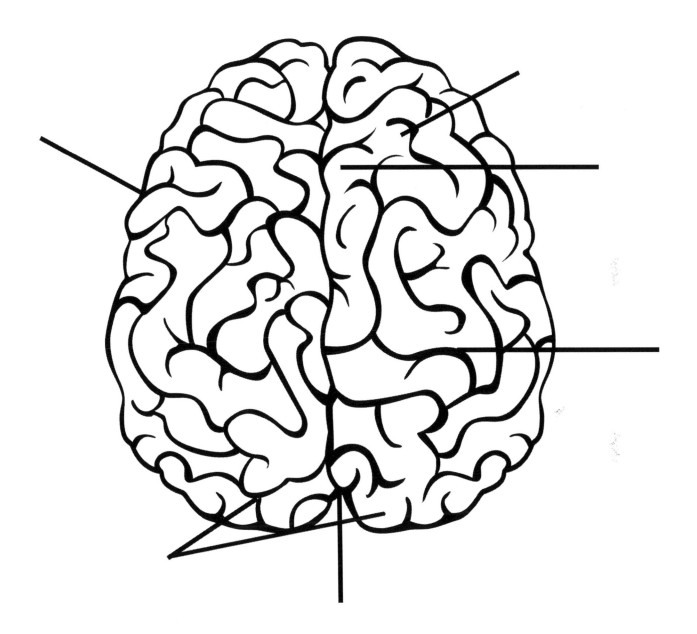

23

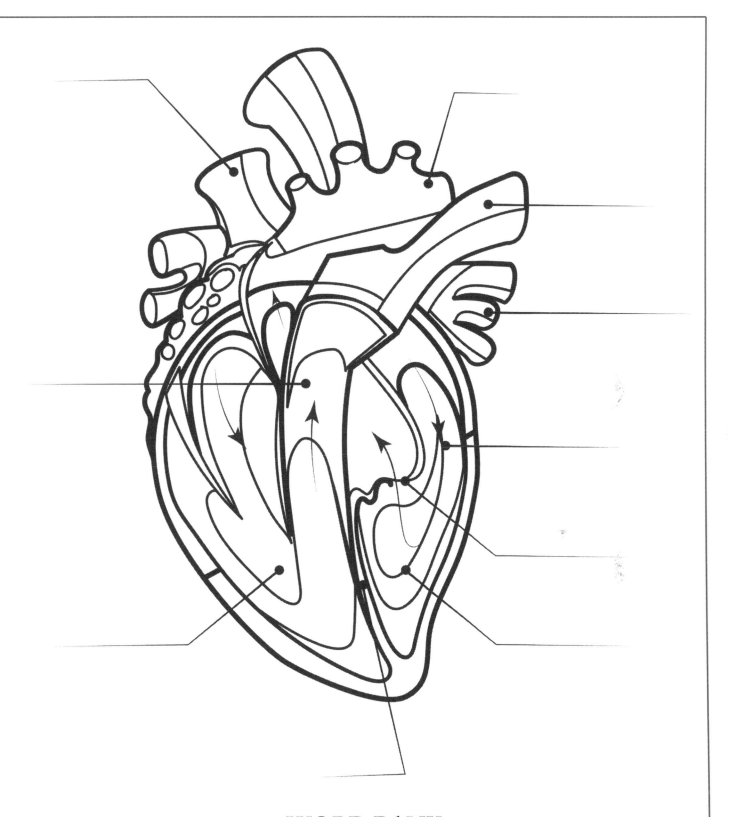

WORD BANK

Septum

Aortic valve

Left ventricle

Right ventricle

Superior vena cava

Right ventricle

Aorta

Mitral valve

Pulmonary vein

Pulmonary valve

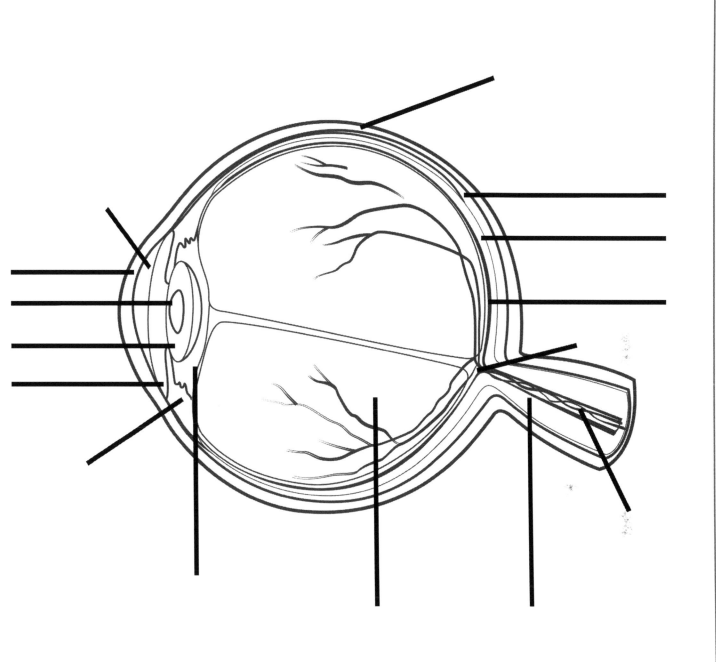

WORD BANK

Cornea	Ciliary body	Anterior chamber aqueous humour	
Iris	Suspensory ligament	Choroid	Sclera
Pupil	Vitreous body	Retina	Lens
Optic nerve	Optic disc blind spot	Fovea	Blood vessels

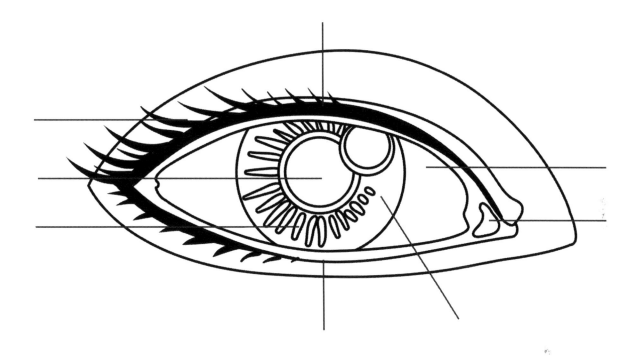

WORD BANK

Iris	Sclera	Lower eyelid
Eyelash	Tear duct	Cornea
Upper eyelid	Pupil	

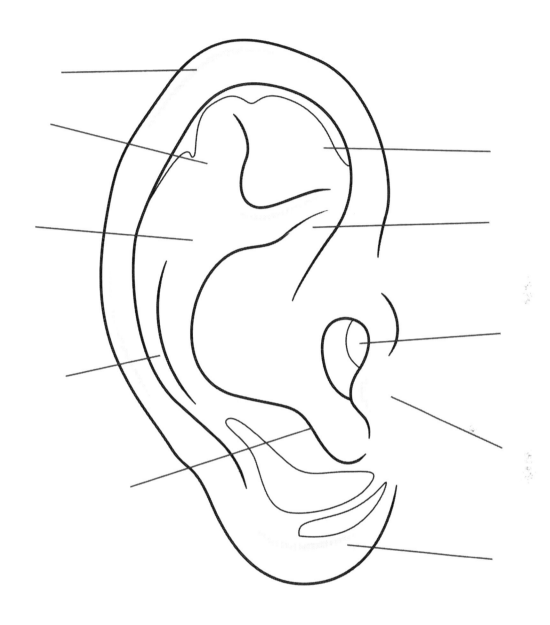

WORD BANK

Helix	Antihelical fold	Antitragus
Scapha	Antihelix	Lobule
External auditory meatus	Tragus	Concha
Fossa		

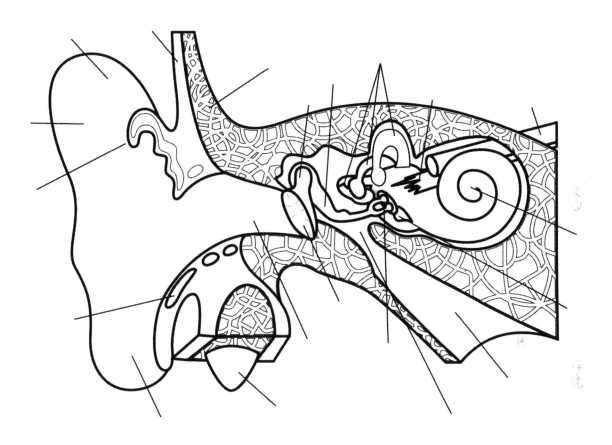

WORD BANK

Mastoid process	Tympanic cavity	Facial nerve	Temporal muscle
Auricular lobule	Stapes	Cochlear nerve	Temporal bone
Cartilage	Ear Drum	Cochlea	Malleus
Tringular fossa	Ear canal	Round window	Incus
Scapha	Helix	Eustachian tube	Semicircular canals

WORD BANK

Cribiform plate	Frontal sinus	Vestibule
Sphenoid sinus	Middle turbinate	Nasal cavity
Sella turcica	Superior turbinate	Inferior turbinate
Choana		

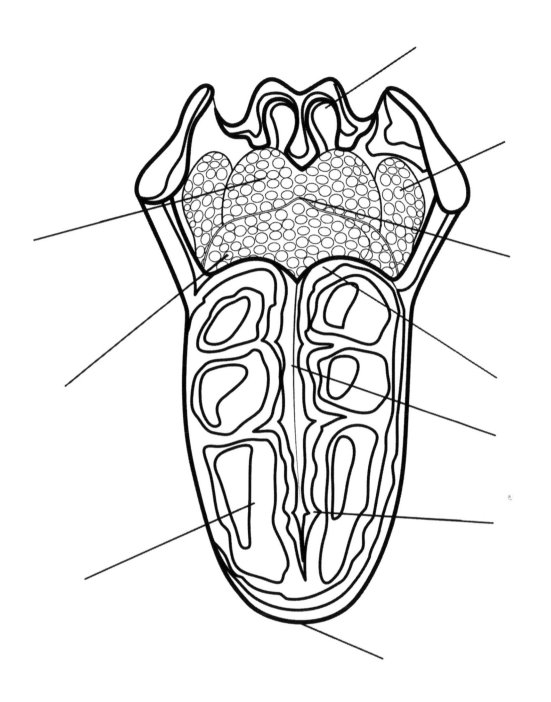

WORD BANK

Apex

Fungiform lingual papillae

Median lingual sulcus

Sulcum terminalis

Foramen cecum

Palatine tonsil

Epiglottis

Filliform lingual papillae

Circum lingual papillae

Lingual tonsil

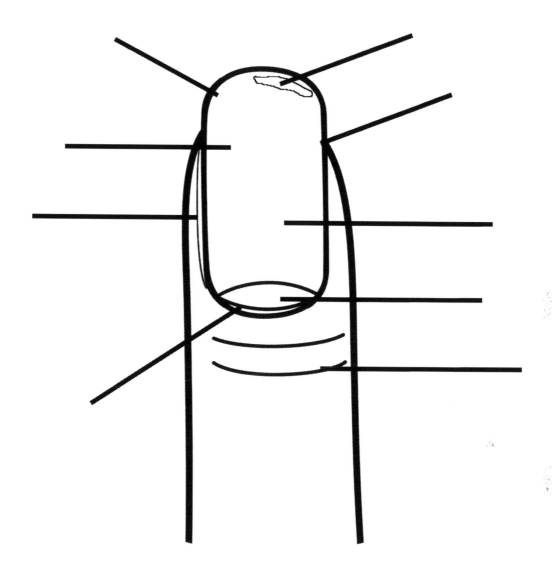

WORD BANK

Free edge	Cuticle	Nail bed
Nail plate	Lunula	Nail grooves
Mantle	Hyponychium	Nail wall

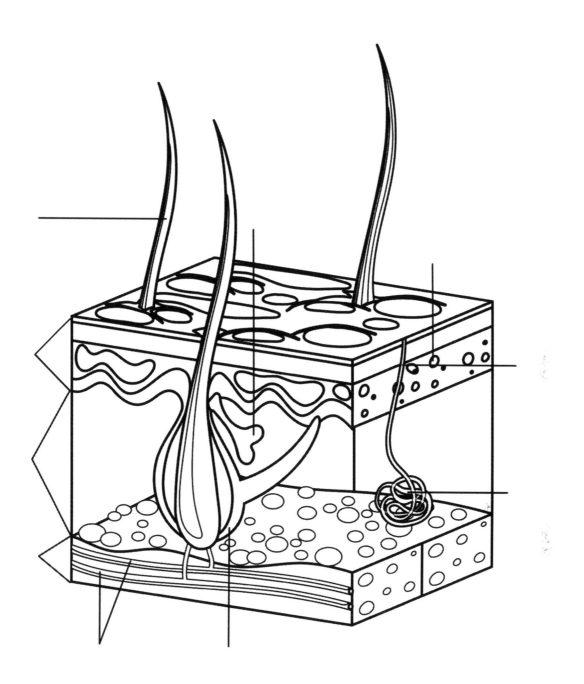

WORD BANK

Sweat gland

Melanocytes

Follicle

Oil gland

Blood vessels

Epidermis

Dermis

Sweat pore

Hair

Fatty tissue

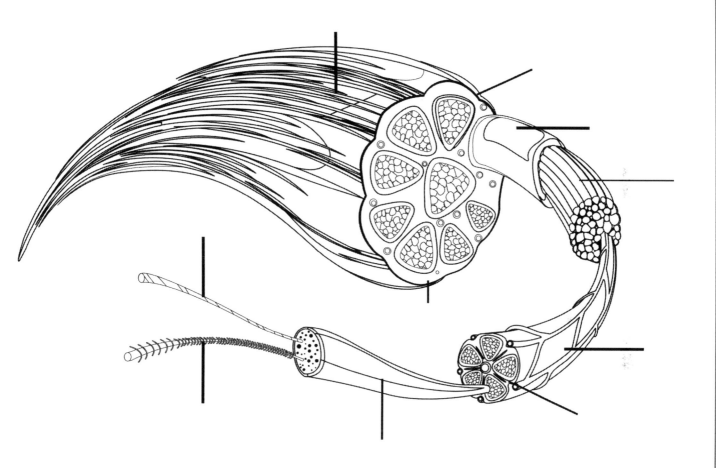

WORD BANK

Myosin	Sarcolemma	Perimysium
Actin thin filament	Muscle fascicles	Saracoplasm
Skeletal muscle	Epimysium	Fasciculus
Muscle fiber		

43

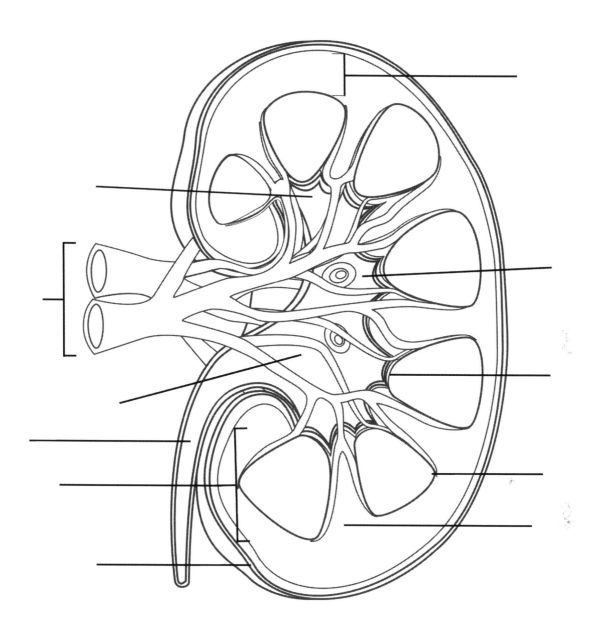

WORD BANK

Minor calyx	Cortex	Renal column
Capsule	Medulla	Renal artery and vein
Ureter	Papilla	Renal pevis
Pyramid	Major calyx	

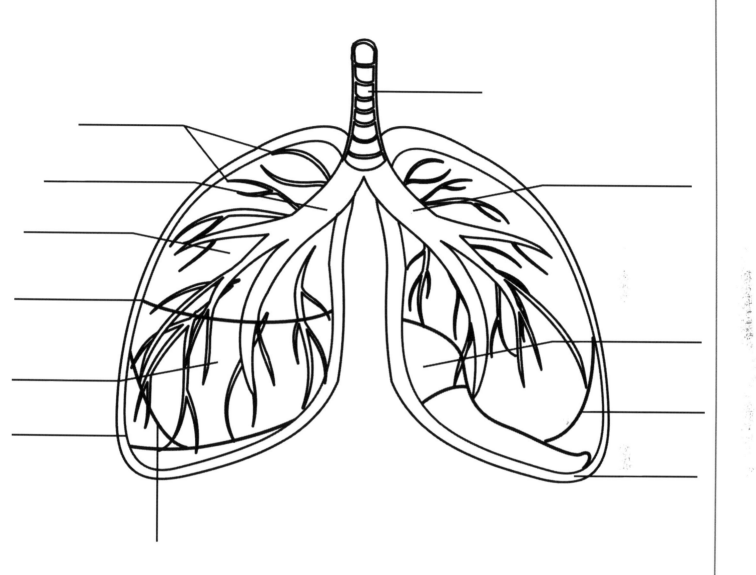

WORD BANK

Pleura	Bronchioles	Right primary bronchus
Carduac notch	Superior lobe	Oblique fissure
Oblique fissure	Horizontal fissure	Trachea
Left Primary bronchus	Middle lobe	Inferior lobe

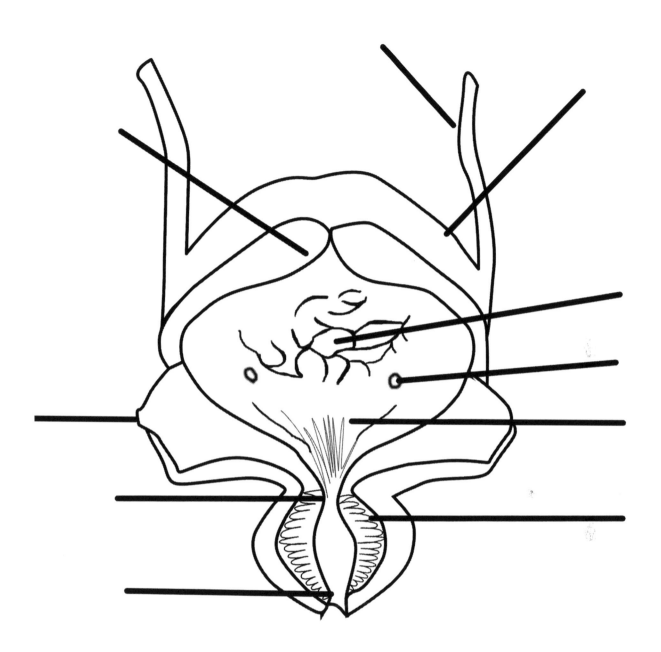

WORD BANK

Destrusor muscle	Trigone	Ureter
Fibrous connective tissue	Prostate gland	Peritoneum
Internal urethral orifice	External urethral orifice	Rugae
Ureteral opening		

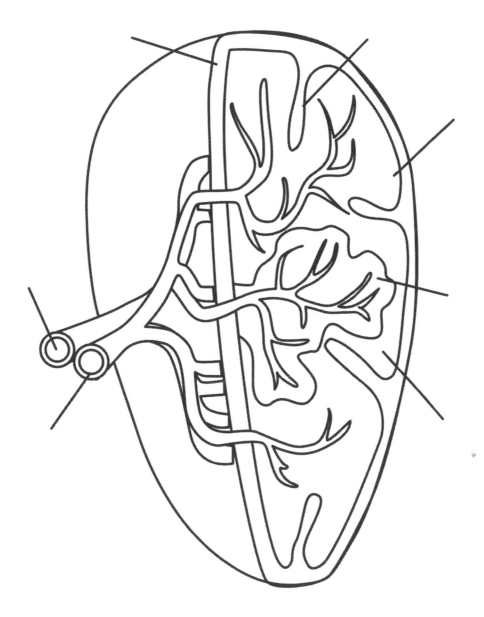

WORD BANK

Vein

Artery

Trabecula

Red pulp

White pulp

Capsule

Vascular sinusoid

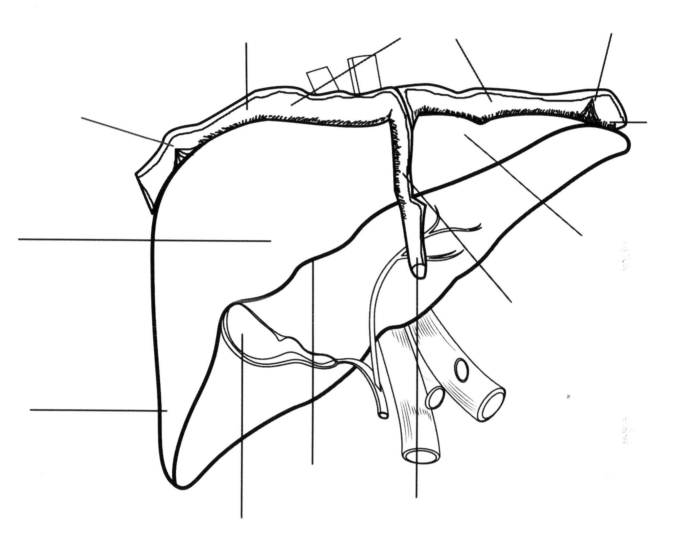

WORD BANK

Fibrous appendix	Right triangular ligament	Costel impressions
Falciform ligament	Diaphragm	Right lobe of liver
Left lobe of liver	Coronary ligament	Gallbladder
Round ligament of liver	Left triangular ligament	Inferior border of liver

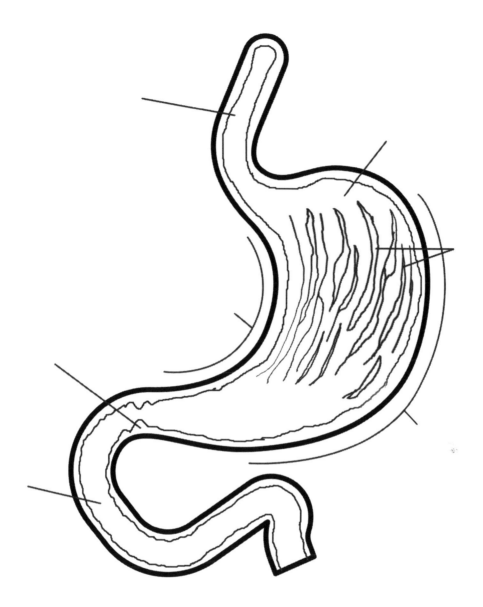

WORD BANK

Rugae

Fundus

Greater curvature

Pyloric sphincter

Circular folds

Esophagus

Lesser curvature

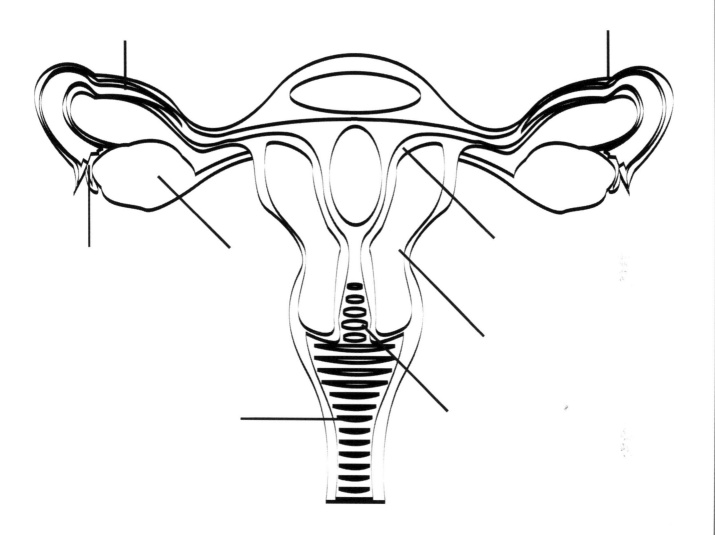

WORD BANK

Cervix	Ovary	Fallopian tube
Myometrium	Vagina	Fimbriae
Endometrium		

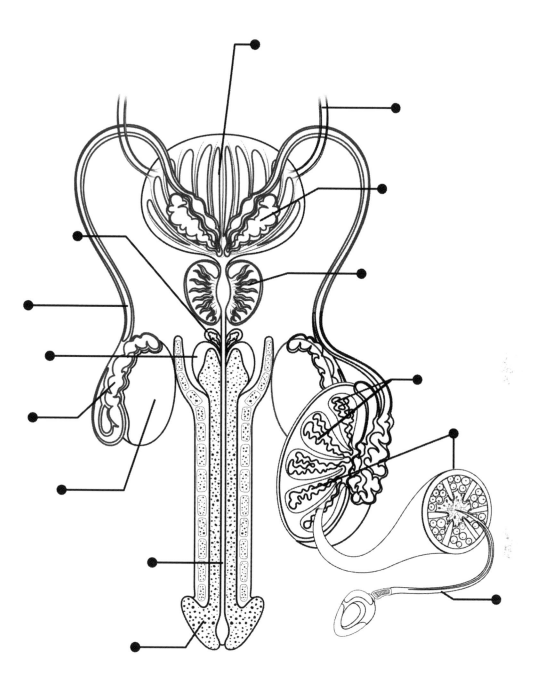

WORD BANK

Urinary bladder	Seminal vesicle	Prostate gland	Bulbourethral gland
Ureter	Urethra	Sperm	Ductus deferens
Sperm-producing tubes	Lobules	Glans of penis	Bulb
Testis	Epididymis		

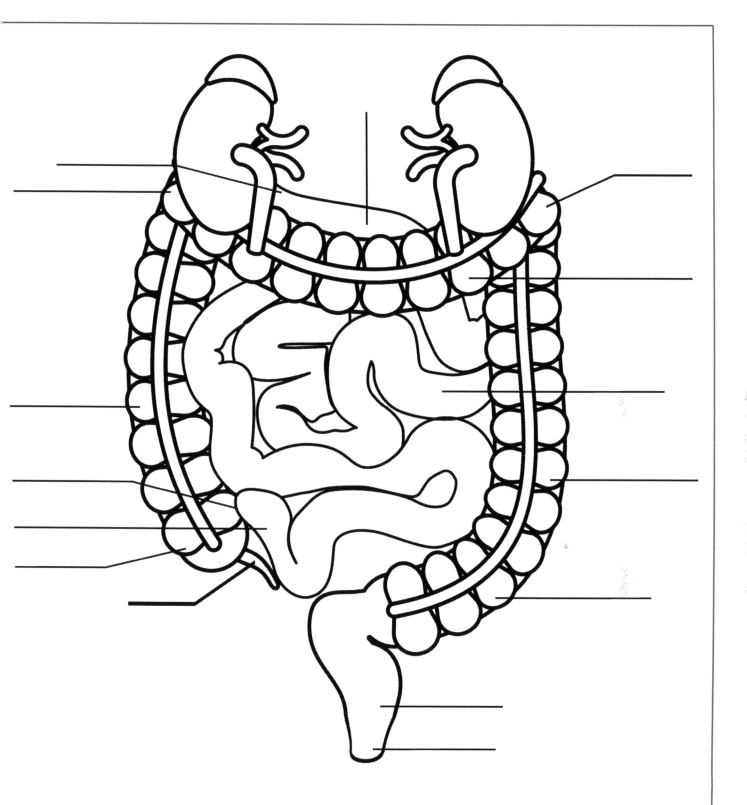

WORD BANK

Left colic flexure	Appendix	Sigmoid colon
Transverse colon	Cecum	Rectum
Jejunum	Ileum	Anal canal
Ascending colon	Duodenum	Descending colon
Duodenojejunal junction	Right colic flexure	Ileocecal junction

61

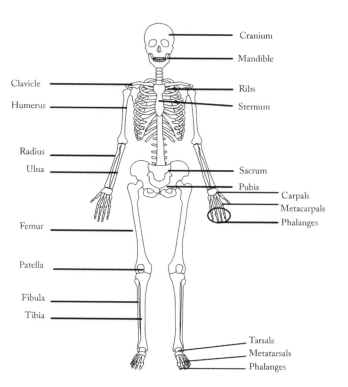

Cranium

Mandible

Clavicle

Ribs

Humerus

Sternum

Radius

Ulna

Sacrum

Pubis

Carpals

Metacarpals

Phalanges

Femur

Patella

Fibula

Tibia

Tarsals

Metatarsals

Phalanges

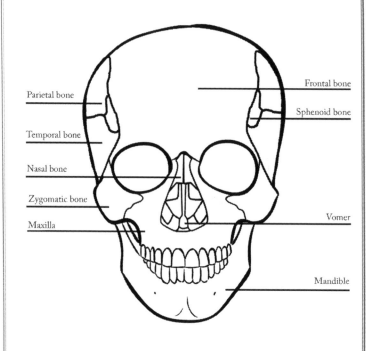

Parietal bone

Frontal bone

Sphenoid bone

Temporal bone

Nasal bone

Zygomatic bone

Maxilla

Vomer

Mandible

WORD BANK

Clavicle	Tarsals	Sacrum	Cranium	Femur
Humerus	Metatarsals	Carpals	Mandible	Patella
Radius	Phalanges	Metacarpals	Sternum	Tibia
Ulna	Pubis	Phalanges	Ribs	Fibula

3

WORD BANK

Frontal bone	Temporal bone	Vomer
Parietal bone	Nasal bone	Maxilla
Sphenoid bone	Zygomatic bone	Mandible

5

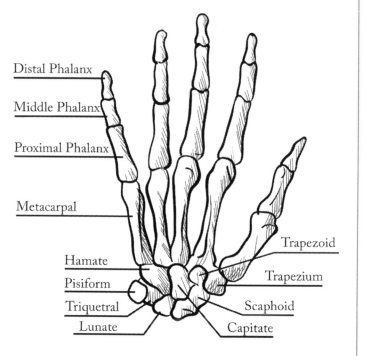

Distal Phalanx

Middle Phalanx

Proximal Phalanx

Metacarpal

Hamate

Pisiform

Triquetral

Lunate

Trapezoid

Trapezium

Scaphoid

Capitate

Cavicle

Scapula

Humerus

Radius

Ulna

WORD BANK

Distal Phalanx	Triquetral	Metacarpal
Middle Phalanx	Lunate	Hamate
Trapezium	Trapezoid	Pisiform
Scaphoid	Proximal Phalanx	Capitate

7

WORD BANK

| Clavicle | Humerus | Radius |
| Scapula | Ulna | |

9

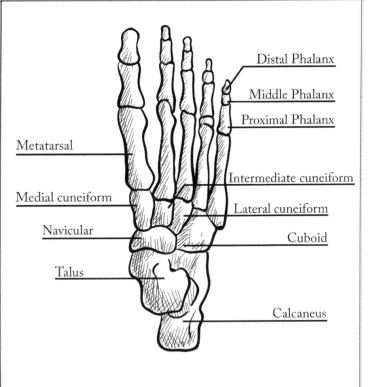

Distal Phalanx

Middle Phalanx

Proximal Phalanx

Metatarsal

Intermediate cuneiform

Medial cuneiform

Lateral cuneiform

Navicular

Cuboid

Talus

Calcaneus

WORD BANK

Distal Phalanx	Medial cuneiform	Calcaneus
Middle Phalanx	Intermediate cuneiform	Cuboid
Proximal Phalanx	Lateral cuneiform	Navicular
Metatarsal	Talus	

11

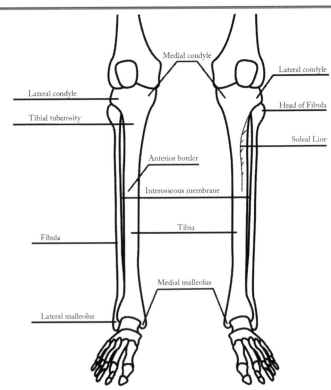

Medial condyle

Lateral condyle

Lateral condyle

Head of Fibula

Tibial tuberosity

Soleal Line

Anterior border

Interosseous membrane

Fibula

Tibia

Medial malleolus

Lateral malleolus

WORD BANK

Lateral condyle	Fibula	Articular surface
Medial condyle	Lateral malleolus	Head of Fibula
Anterior border	Tibia	Tibial tuberosity
Interosseous membrane	Soleal Line	Lateral condyle

1.

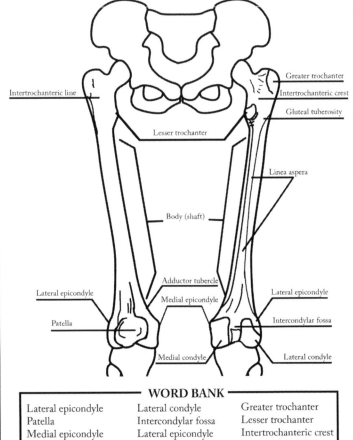

Greater trochanter

Intertrochanteric line

Intertrochanteric crest

Gluteal tuberosity

Lesser trochanter

Linea aspera

Body (shaft)

Adductor tubercle

Lateral epicondyle

Medial epicondyle

Lateral epicondyle

Patella

Intercondylar fossa

Medial condyle

Lateral condyle

WORD BANK

Lateral epicondyle	Lateral condyle	Greater trochanter
Patella	Intercondylar fossa	Lesser trochanter
Medial epicondyle	Lateral epicondyle	Intertrochanteric crest
Adductor tubercle	Linea aspera	Intertrochanteric line
Medial condyle	Gluteal tuberosity	Body (shaft)

15

Iliac fossa

Ant. superior iliac spin

Iliac crest

Iliun

Sciatic notch

Sacrotuberous ligame

Sacrospinous ligame

Ant.
inferior iliac spin

Pubis

Obturator foramen

Pubic tubercle

Ischium

Pubic symphysis

WORD BANK

Sciatic notch	Pubic symphysis	Iliac fossa
Ant. superior iliac spin	Pubis	Iliac crest
Ant. inferior iliac spin	Ischium	Sacrotuberous ligament
Pubic tubercle	Obturator foramen	Sacrospinous ligament
Ilium		

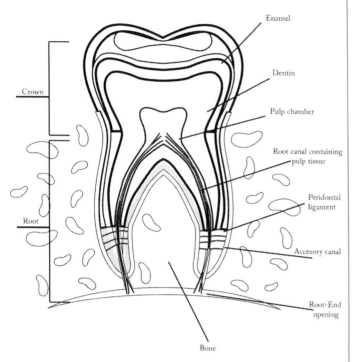

Crown

Root

Enamel

Dentin

Pulp chamber

Root canal containing pulp tissue

Peridontal ligament

Accesory canal

Root-End opening

Bone

WORD BANK

Bone	Dentin	Root
Accesory canal	Enamel	Peridontal ligament
Root-End opening	Pulp chamber	Crown
Root canal containing pulp tissue		

19

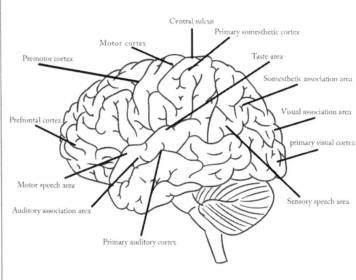

Central sulcus

Motor cortex

Premotor cortex

Prefrontal cortex

Motor speech area

Auditory association area

Primary auditory cortex

Primary somesthetic cortex

Taste area

Somesthetic association area

Visual association area

primary visual cortex

Sensory speech area

WORD BANK

Motor cortex	Motor speech area	Primary somesthetic cortex
Primary auditory cortex	Taste area	Prefrontal cortex
Primary visual cortex	Central sulcus	Premotor cortex
Somesthetic association area	Auditory association area	Sensory speech area
Visual association area		

21

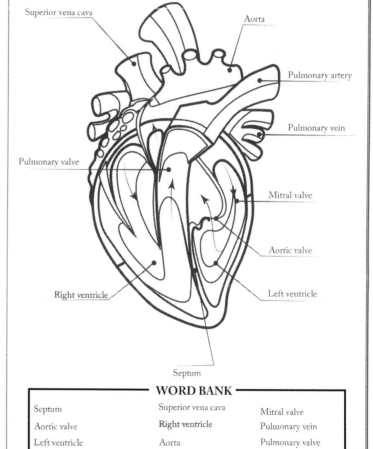

Lateral sulcus

Occipital gyri

Middle frontal gyrus

Superior frontal gyrus

Intra parietal sulcus

Medial Longitudinal fissure

WORD BANK

Occipital gyri	Lateral sulcus
Intra parietal sulcus	Medial Longitudinal fissure
Middle frontal gyrus	Superior frontal gyrus

23

Superior vena cava

Aorta

Pulmonary artery

Pulmonary vein

Pulmonary valve

Mitral valve

Aortic valve

Right ventricle

Left ventricle

Septum

WORD BANK

Septum	Superior vena cava	Mitral valve
Aortic valve	Right ventricle	Pulmonary vein
Left ventricle	Aorta	Pulmonary valve
Right ventricle		

65

25

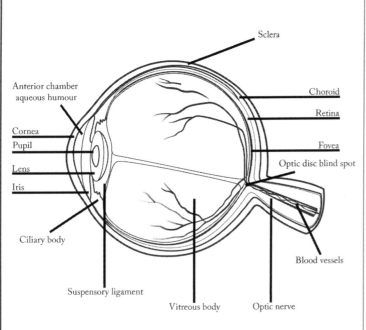

Sclera

Anterior chamber
aqueous humour

Choroid

Retina

Cornea

Fovea

Pupil

Optic disc blind spot

Lens

Iris

Ciliary body

Blood vessels

Suspensory ligament

Vitreous body

Optic nerve

WORD BANK

Cornea	Ciliary body	Anterior chamber aqueous humour	
Iris	Suspensory ligament	Choroid	Sclera
Pupil	Vitreous body	Retina	Lens
Optic nerve	Optic disc blind spot	Fovea	Blood vessels

27

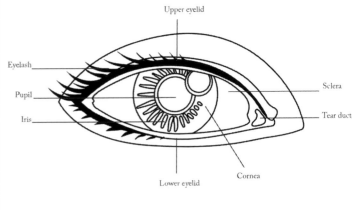

Upper eyelid

Eyelash

Sclera

Pupil

Tear duct

Iris

Cornea

Lower eyelid

WORD BANK

Iris	Sclera	Lower eyelid
Eyelash	Tear duct	Cornea
Upper eyelid	Pupil	

29

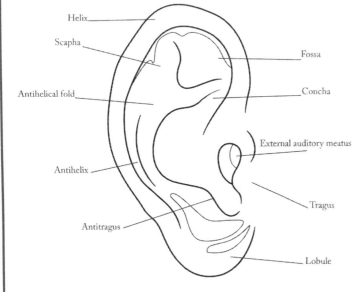

Helix

Scapha

Fossa

Concha

Antihelical fold

External auditory meatus

Antihelix

Tragus

Antitragus

Lobule

WORD BANK

Helix	Antihelical fold	Antitragus
Scapha	Antihelix	Lobule
External auditory meatus	Tragus	Concha
Fossa		

31

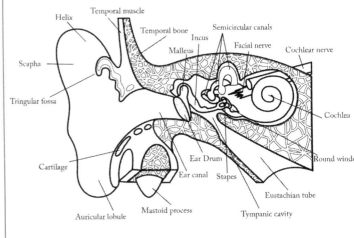

Helix

Temporal muscle

Temporal bone

Semicircular canals

Incus

Facial nerve

Malleus

Cochlear nerve

Scapha

Tringular fossa

Cochlea

Cartilage

Round window

Ear Drum

Ear canal

Stapes

Eustachian tube

Auricular lobule

Mastoid process

Tympanic cavity

WORD BANK

Mastoid process	Tympanic cavity	Facial nerve	Temporal muscle
Auricular lobule	Stapes	Cochlear nerve	Temporal bone
Cartilage	Ear Drum	Cochlea	Malleus
Tringular fossa	Ear canal	Round window	Incus
Scapha	Helix	Eustachian tube	Semicircular canals

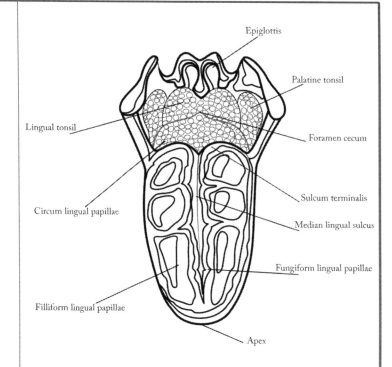

Epiglottis

Palatine tonsil

Lingual tonsil

Foramen cecum

Sulcum terminalis

Median lingual sulcus

Circum lingual papillae

Fungiform lingual papillae

Filliform lingual papillae

Apex

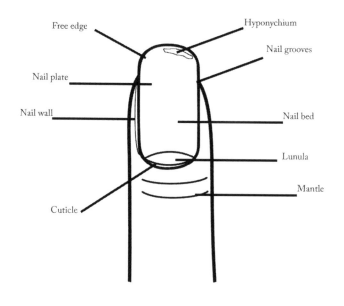

Free edge

Hyponychium

Nail grooves

Nail plate

Nail wall

Nail bed

Lunula

Mantle

Cuticle

Hair

Oil gland

Melanocytes

Epidermis

Sweat pore

Dermis

Sweat gland

Fatty tissue

Blood vessels

Follicle

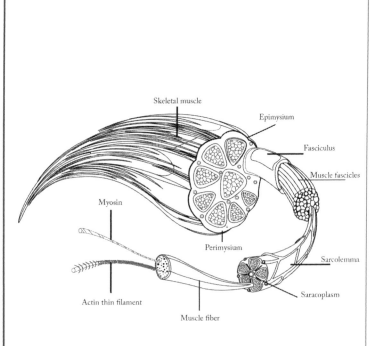

Skeletal muscle

Epimysium

Fasciculus

Muscle fascicles

Myosin

Perimysium

Sarcolemma

Actin thin filament

Saracoplasm

Muscle fiber

WORD BANK

Myosin	Sarcolemma	Perimysium
Actin thin filament	Muscle fascicles	Saracoplasm
Skeletal muscle	Epimysium	Fasciculus
Muscle fiber		

43

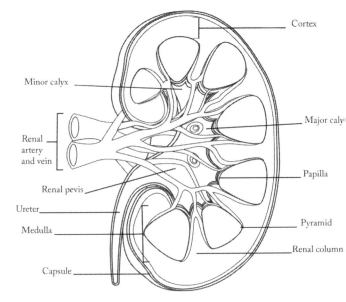

Cortex

Minor calyx

Major caly

Renal artery and vein

Papilla

Renal pevis

Ureter

Medulla

Pyramid

Renal column

Capsule

WORD BANK

Minor calyx	Cortex	Renal column
Capsule	Medulla	Renal artery and vein
Ureter	Papilla	Renal pevis
Pyramid	Major calyx	

4

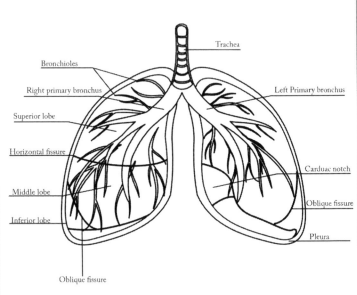

Trachea

Bronchioles

Right primary bronchus

Left Primary bronchus

Superior lobe

Horizontal fissure

Carduac notch

Middle lobe

Oblique fissure

Inferior lobe

Pleura

Oblique fissure

WORD BANK

Pleura	Bronchioles	Right primary bronchus
Carduac notch	Superior lobe	Oblique fissure
Oblique fissure	Horizontal fissure	Trachea
Left Primary bronchus	Middle lobe	Inferior lobe

47

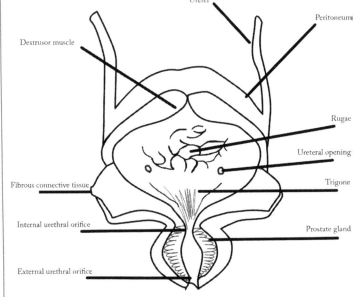

Ureter

Peritoneum

Destrusor muscle

Rugae

Ureteral opening

Fibrous connective tissue

Trigone

Internal urethral orifice

Prostate gland

External urethral orifice

WORD BANK

Destrusor muscle	Trigone	Ureter
Fibrous connective tissue	Prostate gland	Peritoneum
Internal urethral orifice	External urethral orifice	Rugae
Ureteral opening		

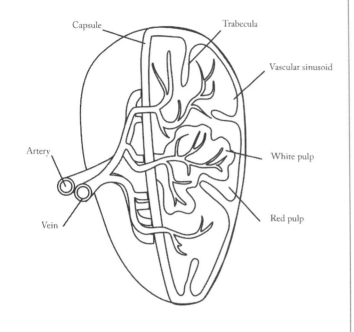

Capsule

Trabecula

Vascular sinusoid

Artery

White pulp

Vein

Red pulp

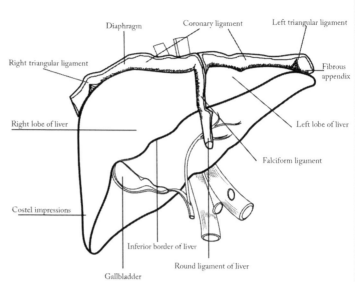

Diaphragm

Coronary ligament

Left triangular ligament

Right triangular ligament

Fibrous appendix

Right lobe of liver

Left lobe of liver

Falciform ligament

Costel impressions

Inferior border of liver

Round ligament of liver

Gallbladder

WORD BANK

Vein	Artery	Trabecula
Red pulp	White pulp	Capsule
Vascular sinusoid		

WORD BANK

Fibrous appendix	Right triangular ligament	Costel impressions
Falciform ligament	Diaphragm	Right lobe of liver
Left lobe of liver	Coronary ligament	Gallbladder
Round ligament of liver	Left triangular ligament	Inferior border of liver

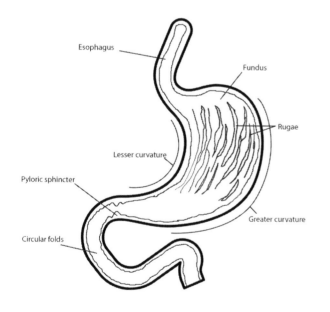

Esophagus

Fundus

Rugae

Lesser curvature

Pyloric sphincter

Greater curvature

Circular folds

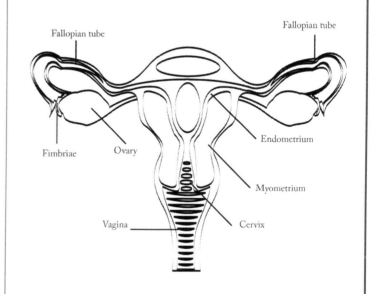

Fallopian tube

Fallopian tube

Endometrium

Fimbriae

Ovary

Myometrium

Vagina

Cervix

WORD BANK

Rugae	Pyloric sphincter	Esophagus
Fundus	Circular folds	Lesser curvature
Greater curvature		

WORD BANK

Cervix	Ovary	Fallopian tube
Myometrium	Vagina	Fimbriae
Endometrium		

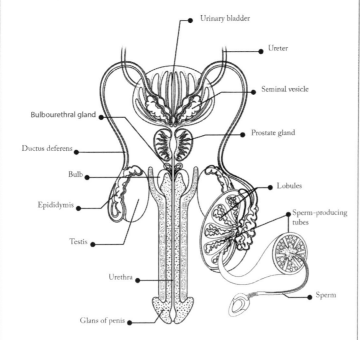

Urinary bladder

Ureter

Seminal vesicle

Bulbourethral gland

Prostate gland

Ductus deferens

Bulb

Lobules

Epididymis

Sperm-producing tubes

Testis

Urethra

Glans of penis

Sperm

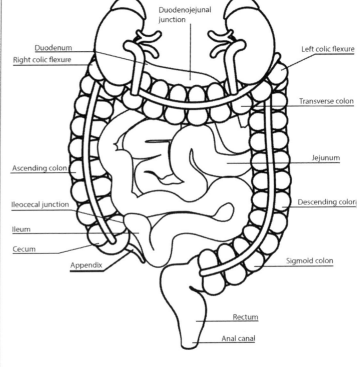

Duodenojejunal junction

Duodenum

Right colic flexure

Left colic flexure

Transverse colon

Ascending colon

Jejunum

Descending colon

Ileocecal junction

Ileum

Cecum

Appendix

Sigmoid colon

Rectum

Anal canal

ANATOMY

Coloring Book for Students & Even Adults

The Anatomy Coloring book and Physiology Workbook with Magnificent Learning Structure

Printed in Great Britain
by Amazon